Contents

TOOL FOR TOMORROW

New Knowledge About Genes

COURTESY NATIONAL INSTITUTES OF HEALTH

by SYLVIA ENGDAHL and
RICK ROBERSON

Atheneum 1979 New York

LIBRARY OF CONGRESS CATALOGING IN PUBLICATION DATA

Engdahl, Sylvia Louise.
Tool for tomorrow.

SUMMARY: Discusses genetic engineering and its applications, including counseling of expectant parents, better utilization of world resources, and improvement of medical care.
1. Genetics—Social aspects—Juvenile literature.
2. Genetics engineering—Social aspects—Juvenile literature. [1. Genetics. 2. Genetic engineering] I. Roberson, Rick, joint author.
II. Title.
QH438.7.E53 301.24'3 78-13777
ISBN 0-689-30679-2

Published simultaneously in Canada by
McClelland & Stewart, Ltd.
Manufactured by American Book, Saddle Brook, New Jersey
Designed by Mary M. Ahern
First Edition

TOOL FOR TOMORROW

New Knowledge About Genes

TOOL FOR TOMORROW

New Knowledge About Genes

Preface

An idea common in science fiction is artificial creation of life forms, or alteration of existing ones, including the human race itself. Recently, a great deal has been written about how this idea is no longer science fiction. Scientists are learning to do some of the things that once could be done only in stories. People have begun to hope—or fear—that science can eventually do even more fantastic things with new discoveries in biology. Such work is generally known as genetic engineering, and it's often said that its goal is to redesign life by unnatural means. Some people call it "playing God." Others feel that even if it is not against God or nature for humanity to alter life forms, the capability will be misused by evil men or will result in disastrous accidents. The most widespread opinion seems to be that whether or not this happens, any genetic engineering will radically change our civilization.

However, the term "genetic engineering" means a lot more than modifying living things in ways that would cause drastic transformations of society. It also refers to a search for significant ways of improving people's health and well-being in a society fundamentally like today's. Not everyone agrees, for instance, that genetic engineering necessarily leads to abolishment of families or to outside interference in each family's decision about having children. Certainly not everyone feels it will be used to alter the characteristics of the human species—I, for one, do not believe it will. But this is something of which those who think of it in science fiction terms aren't always aware.

Because I am known to many readers as a "science fiction author," these readers may assume my own interest in genetics arose from science fiction. That wasn't what happened. My association with science fiction has been centered upon space exploration. I paid scant attention to biology until Rick Roberson, who is a science student, began telling me about current scientific developments in that field. As a matter of fact, I myself had some mistaken notions about genetics when I started my research for this book; I approached the work as a professional writer without previous background in the subject. And as I progressed in that research, I became fascinated by what is happening in genetics. I found that neither fictional speculations nor the facts of genetics traditionally taught—those concerning laws of heredity—give a true picture of its scope.

It's commonly thought that the main function of

genes is to pass along hereditary characteristics. Almost everyone has heard that genes determine whether a person has brown eyes or blue eyes, for example. Most people know that a baby gets one half its genes from its mother and the other half from its father, and that by this means offspring inherit traits from their ancestors. However, though genes do control heredity, whether this should be called their "main" function is a controversial question. Genes control far more than heredity. Virtually every cell of the body contains the same genes as every other; these genes are inherited, but in most cells their function is not to pass traits to descendants. It is to direct the cell in doing its special job in maintaining the body's life. Some people, myself included, believe that maintaining life processes is more fundamental than producing offspring. Personally, I don't agree with the well-known saying that "a hen is only an egg's way of making another egg," although there are scientists who do. In any case, genes regulate all the physical processes of life continuously and not just during an organism's development. The science of genetics has wider applications than most people imagine; it is not something of concern only to those planning to have babies.

Science fiction speculations about genetic engineering—and many of the fears being expressed in today's nonfiction—are focused on the concept of changing existing arrangements of genes to produce creatures with preplanned traits. This is often called "creating new forms of life," although I don't think that's an accurate description. Altering life is not at all the same as creating it from something not already alive. Scientists do not have the ability to create life. What they are gaining is simply the ability to alter some characteristics of living things.

This ability need not necessarily be used to redesign species. It might also be used to correct medical problems, not only in unborn babies, but in older people. All hereditary diseases are genetic diseases, caused by defective genes—but the reverse isn't necessarily true. Things can go wrong with genes after birth, and doctors may someday think of diseases like cancer as "genetic" in the sense that certain genes have been damaged and need repair. Many people find the medical potential of genetic research more exciting than the idea of "creating" life forms (or babies) with planned characteristics. On the other hand, developing altered life forms that can produce more food than today's crops and livestock is indeed an exciting prospect; and it is a far more immediate possibility than redesign of human beings.

Only the last chapter of this book deals with "science fiction type" uses of genetics to bring about new ways of making babies or new physical abilities for humans. And to me, such uses seem less central to the field than the less sensational goals toward which scientists are now working.

—SYLVIA ENGDAHL

Introduction

During the past few years, scientists have become more and more excited about new knowledge of genetics. Once, the word "genetics" meant just the science of heredity. It concerned only how characteristics of living things are passed from one generation to the next by units called genes. However, biologists discovered that genes control all the physical traits and functions of every living thing throughout its lifetime. The science of genetics is now wider than the study of heredity alone. And most scientists feel knowledge in this field can be used in dramatic new ways to help humanity.

What is a gene? Scientists cannot yet give a simple answer to that question. There are many definitions of the word "gene." Some scientists think of genes as units of heredity, for example, while others think of them in terms of the genes' ongoing jobs. Still others describe them in chemical terms. Genes provide biochemical instructions for the development and functioning of each living organism; a great deal about their chemical structure has been learned recently. But much more knowledge is needed before scientists can tell how most genes are put together and how they work. This is especially true of human genes. Every cell of the human body contains hundreds of thousands of genes; so far it has been possible to identify relatively few of them.

Since genes control the body's characteristics and functions, it might seem that different parts of an animal or plant would have different genes. That is not the case. Practically every cell of an individual organism contains the same genes: a complete set of genes for that individual's whole body. Cells in a person's hand, for instance, contain the genetic information for making legs as well as hands. Cells in a hand also contain the genes that tell various glands of the body what to do and how to do it. Why, then, are cells in different parts of the body different? What determines the particular function a given cell will have if it has the same genes as all other cells?

This is another question scientists cannot yet answer completely. They know that in each different cell, some genes are "turned on" and others are "turned off." They know the job of some genes is to control other genes. But they know very little about exactly how genes are controlled, and they are just beginning to learn how to do this in experiments with one-celled organisms. They cannot yet turn genes on or off in human bodies.

One of today's major goals in genetics is to find out

more about how genes work. Scientists see many practical benefits to be gained from such knowledge. And many people, scientists and non-scientists alike, believe that these benefits can help to solve some of Earth's most serious problems.

Using knowledge about genetics for practical purposes is often called genetic engineering. This name was chosen to describe the processes of modifying or synthesizing genes, repairing defective ones, or putting existing genes together in new combinations. Sometimes it is used—as it is in this book—to cover older ways of using knowledge about heredity, too.

The word "engineering" means applying scientific principles to the accomplishment of useful work. But it also has other meanings; for instance, it can mean scheming to control people's acts. Unfortunately, therefore, the term "genetic engineering" is sometimes misleading. Because of the way "engineering" is used in statements that have no connection with science, this term suggests to some people that the aims of genetic engineering are all concerned with controlling people against their will. That is not true. Most goals of genetic engineering have nothing to do with altering human beings, except in the sense of providing better medical care to those who seek it.

Medical science is almost all "engineering," actually. Certainly it all involves application of scientific knowledge to constructive tasks. A surgeon who removes a diseased organ or sets a fractured leg makes a planned, purposeful change in a living body. So does a doctor who prescribes drugs to alter the action of a malfunctioning gland. Future physicians who might use genetic engineering techniques to correct physical malfunctions would not be doing anything fundamentally different to their patients. They would merely be using more advanced scientific knowledge than today's doctors possess.

Another important aim of genetic engineering is to find new agricultural techniques that can help to solve the problem of world hunger. Such techniques can be developed much sooner than most forms of genetic medical therapy, and some people consider them even more promising in terms of benefits to the human race.

To be sure, genetic engineering presents potential dangers as well as potential benefits. Like any other kind of knowledge, knowledge about genetics could be misused. This is what many people fear when they hear the word "engineering" applied to living beings. They are afraid genetic engineers might start creating monsters, or "growing" human beings to suit somebody's special interests, or changing people who don't want to be changed. Also, some of the constructive genetic changes that have been talked about—changes that sound like science fiction—might not turn out well in practice. Many people believe no one could know in advance which ones were wise to attempt. And there would surely be disagreement over just which changes should be called "improvements."

These dangers are real. Every advance in knowledge carries the peril of misuse. Furthermore, advances in genetics involve a type of danger that doesn't arise in most sciences: the danger not only of misuse, but of accidents occurring during necessary experiments that could affect the public.

Dangers, however, can be judged only in connection with goals. There are many goals of genetic engineering that the majority of people favor, besides the ones that are more controversial. The controversies are extremely complicated ones. There are more than two sides to each issue, and it would be almost impossible for any one book, even a very long one, to include all the ideas different people have about the goals and the risks.

This book describes some potential benefits that could come from new knowledge of genetics. It describes some potential dangers, too. However, it is only a starting point. Each reader should investigate more ideas before making up his or her own mind about which goals seem worth pursuing.

1

New Uses for an Old Idea

Since prehistoric times people have been learning about how living things pass traits to their offspring. They have been making practical use of that knowledge, too. In fact, it was with the invention of agriculture—the breeding of animals and plants—that human civilization began. So the idea of controlling the reproduction of life is not new. It is older than recorded history.

Today, the science that explores why living things are like their parents is called *genetics*. This science also deals with how living things develop and how they stay healthy. All practical uses of knowledge about genetics can be called *genetic engineering;* but that is simply a new name for one of the concepts on which human civilization was founded.

Of course, people of the past didn't know many

of the things modern scientists know about genetics. They didn't know, for instance, that living things are made of tiny units called *cells,* and that cells contain *genes,* which control what the cells do. They didn't know that genes are passed along from generation to generation, or that it is the genetic makeup of an organism that makes it like its ancestors. However, they did know that to grow wheat, they must plant wheat seeds, not seeds of some other plant. And they knew that cattle give birth to calves, not lambs. They soon learned how to plan the mating of animals and choose the best and strongest offspring to breed again. They learned how to choose seed to produce good crops. By making these choices, they were selecting *genotypes,* though they weren't aware of it. Selection of genotype—genetic makeup—is the basis of all agriculture as well as of genetic engineering.

The domestic animals and cultivated plants that exist today are nearly all products of such selection. For example, there are many different breeds of horses. This didn't happen by accident; it was the result of careful work. Among the first people to breed horses were the Arabs, thousands of years ago. They knew nothing about genes, and their methods were crude by modern standards. These methods, however, followed principles that are still used in animal breeding.

Horses were very important animals to the Arabs,

useful both in peace and in war. Naturally they wanted the best ones they could have; they learned to pick them for desirable characteristics, like speed and stamina, and to breed these horses with one another. The foal of a selected stallion and mare would often combine the best traits of the two. Such a foal had gotten genes for those traits from its parents, but the Arabs didn't know that; they knew only that their method of selection produced superior horses. By continuing the process for generation after generation, they developed the breed called the Arabian horse.

Crossbreeding, the genetic engineering method the Arabs used with horses, is still an important technique. Racing horses and other show animals are often interbred by owners who hope prize-winning offspring will be born. So are farm animals. Farmers try to develop strains of livestock by this process that will produce more meat or milk.

Over the centuries, much more has been learned about inheritance than was known in ancient times. Breeding animals is not just a matter of picking parents with characteristics that are wanted, for it is not practical to control every mating. A breed, or strain, consists of individuals that can mate with any member of that group and produce offspring with the characteristics common to the group. If two pedigreed collies mate, for example, all the puppies will be collies. But if a collie mates with a dachshund, the puppies will be mongrels. Some of their descendants may look like collies, yet not produce collies if mated with each other. This is because they will have inherited some genes from their dachshund ancestor that they can pass on, even though these genes were not the ones that determined their own characteristics. Their genotype will not be entirely "collie."

Actually, although animals are often called "purebred," there is no such thing as a "pure" genotype. Every individual has a unique combination of genes. However, members of the same breed or strain are of very similar genotypes. If uncontrolled mating outside the group were allowed to occur, characteristics of the offspring would be unpredictable.

For many years now, controlled mating has not been the only way of producing superior livestock. A more efficient way, especially in the case of cattle, is *artificial insemination*. With this method, cows are made pregnant with stored sperm from the best bulls. This means that prize bulls can be the fathers of thousands of calves all over the world, since sperm can be frozen and shipped. It also means not every farmer who wants calves has to keep a bull.

Artificial insemination is now a commonly accepted way of breeding cattle, and is also used with some other kinds of livestock. Furthermore, it is beginning to be used to help preserve species of wild animals that are in danger of extinction. Often there are few such animals in zoos, and shipping males or females from one zoo to another for mating purposes is not always practical. Sometimes zoo animals refuse to mate. With artificial insemination, it is possible for these animals to have offspring.

Everyone knows that both sexes are needed for the reproduction of animals. People who have not studied plants, however, are sometimes not aware that plants also reproduce sexually. Many plants can be reproduced in other ways: from bulbs or cuttings taken from the original plant, for instance. But when plants are grown from seed, sexual reproduction is involved. A female flower must be fertilized with pollen from a male flower for seed to form, though with some kinds of plants, the pollen of one flower can fertilize female parts of the same flower.

There are many ways of improving plant strains by selecting genotypes. One of them, which is similar to crossbreeding in animals, is called *cross-pollination*. Pollen from exceptionally fruitful plants is used to fertilize, or pollinate, the flowers of similar plants. The desirable genes—and therefore desirable characteristics—of both parent plants are transmitted to some of the new plants that grow from the seeds.

In the late 1960s and early 1970s, cross-pollination

was used to develop better strains of wheat and rice. The new strains produced two or three times more food than the old ones did. Farmers began to plant them, and the switch paid off. There were record harvests all over the world. Besides being more economical to raise, the extra food helped to feed undernourished people in many nations. The changeover to improved grains was called the Green Revolution, and it was so successful that one scientist won a Nobel Peace Prize for his work in developing the plants.

Such improved crops can be a mixed blessing, however. As some scientists have pointed out, switching over entirely to the new strains would be like putting all one's eggs in one basket. Farmers used to plant many different kinds of seed; when they use improved seed, they plant only a few kinds. If a disease or insect pest should get out of control in the fields, it might wipe out all the new plants everywhere because they are all so much alike. If the old kinds of seed were still being planted, there would be more varieties of plants in the fields, some of which might be more resistant to the disease or pest than others. These might survive to produce food when others did not. So while Green Revolution grains are being widely used, there is growing interest among agricultural experts in establishing a "seed bank" to maintain a variety of plant genotypes with which to experiment.

Seeds from many varieties of food crop plants are stored in this "gene bank". Scientists believe it is important to preserve crop species that farmers no longer grow. The improved crops now being planted around the world have less genetic diversity than older kinds did. If a new disease or insect pest should wipe out a crop species, seeds from the "bank" could be used to develop a resistant crop that would survive. COURTESY THE ROCKEFELLER FOUNDATION

Further crossing of crop plants is being tried in laboratories because increasing harvest size alone will not provide food for all the undernourished people in the world. Scientists are trying to develop grains and vegetables that are more nutritious. Cereals and vegetables provide many things that are essential to nutrition; they are important parts of a balanced diet. But by themselves, they are not adequate. In many parts of the world, people do not have a balanced diet; often they lack protein because they can't afford foods like meat, fish, eggs and cheese. Perhaps cereals and vegetables rich in protein can be developed so that these other foods won't be needed for health. Scientists are working to breed such plants, but there is still much to be learned about how plants make protein and what genes control the process.

Still another goal in agriculture is to develop plants that can be irrigated with seawater. Most plants can't tolerate the salt in seawater, and the few that normally can aren't good to eat. Yet if seawater could be used on crops, far more land could be irrigated than is now possible; land now wasted could produce food. Recently, a few kinds of barley that tolerate salt have been found. It may be possible to cross salt-tolerant plants with varieties that yield good harvests if the genotypes possessing salt tolerance can be isolated.

Selective breeding of animals and plants is the

oldest type of genetic engineering, and it is still a good means of improving livestock and crops. The larger Earth's population grows, the more need there will be for better agriculture.

2

Improving the Human Race: Some Hopes & Some Fears

The characteristics of plants and animals can be improved by choosing parent organisms of desirable genotypes. Is the same true of human beings? Could the human race be improved by selecting people with the best genes to have children?

As far as biological science goes, it would be possible to influence the number of children born with chosen characteristics. But with human beings, there is a great deal more involved than biological science. Improvement of the human race by selection of genotype is called *eugenics*. There have been supporters of eugenics since science first began to learn about genes. In fact, even in ancient civilizations some men believed that humanity could be improved by selective breeding. Many people, however, believe that for society to attempt this would be a violation of individual rights. Furthermore, not everyone agrees about what human characteristics are "best."

Another reason people worry about eugenics is that the word eugenics, itself, was misused by the Nazis in Germany; this gave the whole concept a bad name. The Nazis wanted to develop a "Master Race" to conquer Earth. Besides adopting dictatorial eugenics policies for eliminating inherited diseases, they tried to "improve" the population according to their own false views of what kind of people were "superior." The worst of their misnamed measures was the extermination of Jews. Of course, elimination of Jews, or any other ethnic group, is not eugenics—it is not "improvement" at all. Nevertheless, the Nazis claimed it was part of their eugenics program, so the word got associated with their evil actions.

The aim of real eugenics is not to manipulate ethnic groups or encourage the traits of one such group at the expense of others. It is concerned with traits that can appear in individuals of any group. For instance, some scientists believe democratic forms of eugenics should be used to reduce the number of people born with genetic diseases. The extent to which most such diseases could be conquered by this means is uncertain, however.

Genetic diseases are caused by defective genes that give faulty instructions to cells of the body. It is not yet known how many diseases are the result of defective genes. Possibly, some diseases not usually considered "inherited" are actually caused by combinations of genes too complicated for scientists to have analyzed so far. Some may also be caused by damage to a person's genes after birth. Today, only two types of diseases are called "genetic": those known to be passed on from one generation to another in families, and those caused by structural abnormalities in the genetic material of the reproductive cells that combine to start a baby. Most diseases of these types are extremely rare. Many of them are serious, however; their victims often die in infancy or are severely handicapped throughout their lives. The majority strike only persons who receive defective genes from both parents.

Attempting to have fewer children born to parents with genetic defects is one kind of eugenics. Attempting to have more children born to parents with desirable traits is another, and it is in some ways more debatable, since scientists do not agree about how large a role genes play in the development of most such traits. For example, supporters of this type of eugenics have traditionally aimed to increase the average intelligence of the population. But there have been many arguments over the question of whether intelligence is determined by heredity or by environment. Now, most scientists feel that the answer is "both." A person's heredity—that is, genotype—interacts with the environment in which he or she lives as that person's mind develops. So there is no general agreement about whether encouraging especially intelligent individuals to have babies would produce more intelligent citizens.

Environment plays a part even in the development of many diseases. Some genetic diseases do not develop if the babies who suffer from them are kept from eating certain foods. Also, some diseases run in families, yet do not affect all the individuals who may have defective genes. Those in whom these diseases develop are thought by many scientists to have genotypes that react badly to particular foods or to other sorts of environmental "trigger" conditions. Someday, doctors may identify many more diseases of this kind than are now known and learn to tell which people should avoid specific conditions in order to stay well. Though some scientists believe society should try to prevent such people from being born, others believe it is better to control environments than to control genotypes.

Differences of opinion such as these cause many people to oppose eugenics altogether. In the early twentieth century, laws were passed in some places to keep persons assumed to possess seriously defective genes from having children, but many of these laws were later repealed or ignored. Today, even people who favor such laws recognize that genetics is much, much more complicated than was once thought. For example, though some books sort people into "brown-eyed" and "blue-eyed" categories to explain how genes received from each parent affect characteristics, scientists now know this is an oversimplification. There are not just two shades of eye color; even as simple a characteristic as that is controlled by several pairs of genes. Only in the case of rare diseases can the likelihood of a child's inheriting specific traits be known.

When a disease is caused by a single defective gene that always produces visible effects, the odds of a child's inheriting that disease are easy to compute. All human genes come in pairs; a baby gets one gene of each pair from its father and the other from its mother. Exact statistical laws tell the probability of the baby getting a pair of genes that will produce a particular characteristic. However, this method of computing odds can't be used in most cases. Not only are most characteristics determined by multiple pairs of genes instead of single pairs, but scientists usually can't tell how many pairs are involved. Furthermore, if the two genes of a pair are not alike, sometimes only one of the two produces detectable effects. It is not yet possible for doctors to identify the specific genes a person has by ana-

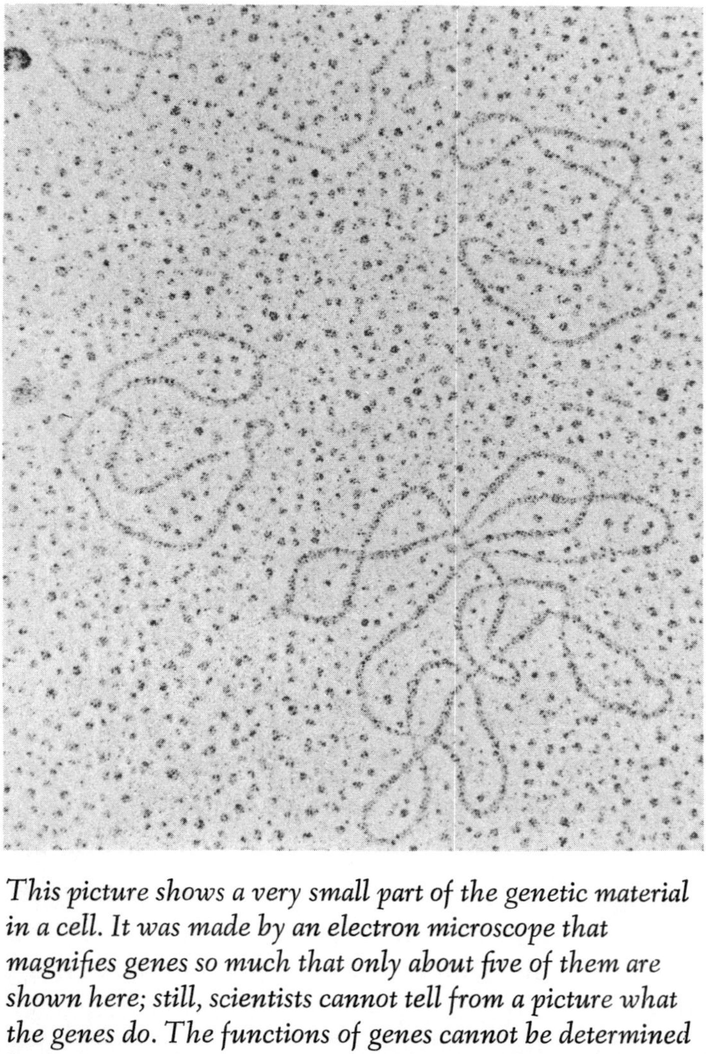

This picture shows a very small part of the genetic material in a cell. It was made by an electron microscope that magnifies genes so much that only about five of them are shown here; still, scientists cannot tell from a picture what the genes do. The functions of genes cannot be determined by looking at them. Even if it could, it would not be possible to make pictures of a person's genes, because each human cell contains hundreds of thousands of genes.

COURTESY BROOKHAVEN NATIONAL LABORATORY

lyzing genetic material itself; every human cell contains hundreds of thousands of genes. Tests can show only the effects these genes produce. Estimating what traits will be passed on to babies is therefore very complicated.

Of course, if both parents have an inherited disease, it is known that they have genes that cause it. But with most such diseases it is possible to have the genes without being sick; the defective genes may be members of pairs in which healthy genes dominate. People who have "hidden," or *recessive,* genes for a disease are said to be *carriers* of that disease. They cannot get the disease, but they can pass it on to their descendants. Usually their children cannot get the disease unless both the father and mother are carriers, though the children themselves can become carriers in turn.

There are still people who believe that individuals with genetic diseases, and even known carriers of such diseases, should be prevented by law from having children. This would reduce the number of defective genes in the human race, but most people would consider it a violation of individual rights. Moreover, there is no agreement on the degree to which it would benefit society. Since the majority of inherited diseases cannot appear in a carrier's children unless his or her mate is another carrier of the same disease, and since these diseases are rare, the chances of such a person producing diseased children are usually quite small. Even in a marriage between carriers, the odds are against any one particular child being afflicted; so forbidding carriers to have children would prevent the birth of healthy babies as well as sick ones. And although some of these healthy babies would be new carriers, not everyone feels that this is necessarily something to avoid.

In some situations a carrier of a disease can be better off than a non-carrier. For example, sickle-cell anemia is a serious disease that affects mainly black people. Those who have the disease often die. Carriers, however, are not diseased; and they are better able to survive the contagious disease malaria than people who don't have a recessive sickle-cell anemia gene. This is of little advantage in the United States, which has an environment almost free of malaria. But in Africa, malaria is a common disease that kills many people, so in Africa being a carrier of sickle-cell anemia is helpful. Would it be good for the human race if all genes for sickle-cell anemia were eliminated? People in the United States are apt to think so, since it would mean no more black children would die of that disease. For people in Africa, where blacks get malaria oftener than they get sickle-cell anemia, the question is more difficult.

Scientists do not yet know how many defective genes can be beneficial in special circumstances. Not all such genes can; some are slightly damaging even to carriers. There may be others, however, that help more people than they harm. Also, some seemingly defective genes might prove helpful if environmental conditions were to change in some unforeseen way. Some scientists feel that keeping as wide a variety of genes as possible available is better for future generations than getting rid of all genes now considered "defective," because variety of genotype is what makes the human race adaptable.

Furthermore, if genetic diseases can be treated, as many now can, is there any reason to prevent the birth of people with treatable defects? The answer to that, too, is debatable. Sometimes the treatment is expensive; and even apart from the cost, many scientists believe that it is bad to have a large number of people dependent on continuous medical care. Until recently, most of these people would have died. Now that more and more treatments that prolong life are being found, the percentage of the population dependent on medicine is increasing; and if the people being treated pass on their defective genes by having children, it will continue to rise. Some scientists fear that in time almost everyone will need treatment of one kind or another. This, they argue, would weaken the human race. Others believe that before that could happen, it will become possible to cure genetic diseases instead of merely treat them.

Still another question in eugenics is whether people's health should be considered the most important factor in evaluating their genotypes. Some genetic diseases do not prevent their victims from making great contributions to society. Many highly talented persons have had such diseases. There is little agreement about the extent to which genes determine talents, but it is generally believed that they do have a role. A common feeling in most societies has been that especially talented people should be encouraged to have children. If every person with any defective genes were prevented from passing them on, his or her other genes would not be passed on either. There is no way of knowing what might be lost.

All of these questions are extremely complex ones. Even if there were scientific agreement about eugenics, there would still be disagreement about the social and ethical problems it involves. In addition, many people who think eugenics might be a good thing in theory worry about governments or other groups using it to control citizens unjustly.

This is not an idle worry. The Nazis did impose dictatorial eugenics policies, and if they had had more scientific knowledge of genetics, their abuses might have been even worse. However, there are a great many ways in which the Nazis might have misused any type of new knowledge. All scientific advances can be used for evil as well as for good.

Today, what genes will be passed to future generations depends on individual choice; couples can decide for themselves whether or not to produce offspring. Progress in the field of genetics is helping more and more couples to make informed choices.

3

Healthier Babies Through Genetic Counseling

The majority of citizens oppose interference by society in anyone's choice of marriage partner or choice about whether or not to have children. However, most feel that people likely to have children with genetic diseases should consider their decisions very carefully. Persons with reasons to suspect they may carry defective genes often worry about their future children's health. Therefore, those with a family history of inherited disease can be helped by genetic counseling, which is advice given to such persons by doctors who have studied genetics. Sometimes individuals seek counseling before deciding whom to marry. More frequently, couples ask for advice after marriage about the risks of having babies.

Many times scientists can compute the likelihood of a genetic disease appearing in a child of a couple, though it is rarely possible to predict whether it will appear in any particular baby. For instance, if both the man and the woman are known to be carriers of a recessive gene for the same disease, the odds are three to one that their baby won't have that disease. There is a fifty-fifty chance that it will be a carrier like the parents. Of course, this is only an example. Most cases are far more complicated than this, both because many diseases involve more than one gene pair and because doctors cannot always determine whether or not the parents are carriers.

Until quite recently, possible carriers of genetic diseases could be identified only by knowing what people had relatives and ancestors with genetic diseases. Usually such people could be told only the probability that they were carriers; in most cases the only sure evidence was their having had a diseased child. Many books say that recessive genes—defective or otherwise—produce no effects in the people who have them. It is still true that the effects of some such genes cannot be detected in carriers. However, it has been found that carriers of certain diseases do have different body chemistries from non-carriers, though not serious enough differences to produce the disease. More and more defects are being added to the list of those for which carriers can be detected through blood tests or other medical procedures. As people become aware of this, it is likely that increasing numbers of couples will seek genetic counseling.

At present, genetic counseling is not used routinely. It is fairly expensive; there are not enough trained counselors and laboratories to counsel everyone who might benefit. The special tests needed to identify carriers are complicated ones that relatively few doctors are equipped to perform. Furthermore, comparatively few people realize that genetic counseling is available. Most of those who request it have relatives with serious genetic diseases or a child afflicted by such a disease. They want to know the odds of their bringing another sick child into the world. If they learn that the chances of their having a healthy baby are good, a great burden is lifted from their minds. If the odds against their producing healthy babies turn out to be great, however, and they decide not to have offspring, they and the diseased children they might have had are spared many problems.

The actual decision about having a baby is made by the prospective parents. The genetic counselor merely provides information about chances that the baby would inherit defective genes, and about the consequences of the disease it might be born with. Though most couples don't wish to run a large risk of having a child with a disease that would cause great suffering or early death, many are willing to risk having a child with a treatable genetic disease. People's views differ widely. Some think it is wrong to add a person who needs treatment to the population, while others feel that as long as they love and care for the child, its illness does not matter. Some believe every child has a right to be born without physical handicaps; to others, such handicaps seem of small significance in judging a person's worth to himself and to society. Naturally, opinions like these depend a great deal on the seriousness of the handicap. At present, scientists can provide more information about rare, serious genetic diseases than about less serious ones. As more is learned about the genetic basis of comparatively minor defects, such decisions may become more difficult.

Many carriers who have genetic counseling do decide to become parents. In fact, many who have been afraid to have children of their own decide to go ahead when they are told by a counselor that the odds in their case are not as poor as they had feared. Genetic counseling is not a form of eugenics. Like other doctors, genetic counselors put the welfare of their patients first, rather than the welfare of the human race as a whole.

However, there are controversial social issues connected with genetic counseling. Some people think that if a carrier is found, his or her brothers, sisters and perhaps even cousins should also be tested by health authorities. Some even think that all couples should be required to have genetic counseling before marriage, or at least before starting a family. Other people believe that health officials have no right to investigate the genetic makeup of anyone who has not sought counseling voluntarily.

Although genetic counselors can rarely be sure before a mother becomes pregnant whether or not her baby will have defective genes, they can sometimes tell after she is pregnant. Unborn babies, called *fetuses*, can be tested for certain genetic diseases. A relatively new medical technique, *amniocentesis*, is used to do this.

During pregnancy, a fetus is surrounded by amniotic fluid produced by the mother. This fluid helps to cushion and protect the fetus during development, and a few skin cells from the fetus will usually be found floating in it. Amniocentesis involves using a hypodermic needle to draw off a sample of this liquid. The risk of hurting the mother or the fetus by the procedure is small if it is done by a doctor with special training, though most doctors believe only mothers with reason to suspect their babies may be diseased should undergo it. This includes all mothers thirty-five or older, since certain types of genetic diseases can be caused by a mother's age even if her own genes are not defective.

After amniocentesis, the fetal skin cells in the sample are grown until enough exist to be tested, and if the tests show that the fetus does have a genetic defect, the pregnancy can be terminated by abortion. Of course, many people do not believe in abortion, even if the fetus has defective genes; they believe that destroying a fetus is the same as murdering a person. Others believe that a baby must be born before it becomes a full human being with human rights apart from the mother's. Still others believe that the fetus becomes entitled to separate rights at some particular stage of pregnancy.

The decision to abort a fetus with a genetic defect is made by the parents, not by doctors or genetic counselors. Some parents do have the child even if amniocentesis shows it will be diseased, although doctors do not usually perform the procedure on mothers who are opposed to the idea of abortion. Amniocentesis saves many fetuses that would otherwise have been aborted because of their parents' fear of genetic defects. Before the technique was developed, carriers often terminated unplanned pregnancies. Now, if amniocentesis proves their baby has no detectable defect, these parents can look forward to a healthy child.

Despite genetic counseling, a great many children are born with unexpected genetic defects. This is partly because not everyone can get counseling, and because

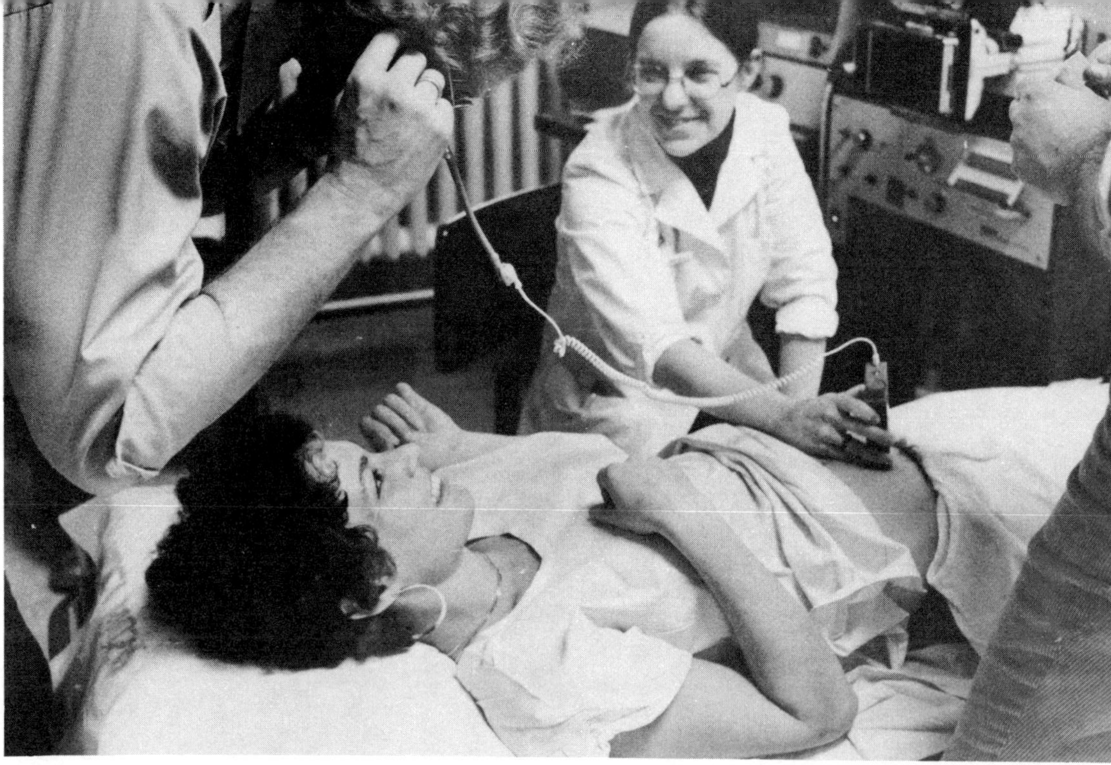

This mother is about to have amniocentesis for reassurance that her unborn baby is healthy. In the picture, her husband is listening to the baby's heartbeat. JIM VINCENT, THE OREGONIAN

some people who don't know their family histories, or who don't have many close relatives, have no reason to suspect they may be carriers. But it is also possible for a baby to have a genetic defect not present in its ancestors. Normally, a baby's genes are exact copies of half the father's genes and half the mother's genes; its genotype is a unique combination of genes that have been passed along for many generations. In very rare instances, however, these genes are changed by damage to a parent's reproductive cells, caused, for example, by radiation or by certain drugs. Such changes are called *mutations*. Mutations are not always harmful; in fact it was through a slow series of mutations, occurring over millions of years, that the various life forms on Earth evolved. Harmful mutations far outnumber helpful ones, however. In addition, genetic defects in children occasionally result from something going wrong during the process of reproduction. Therefore, even if all couples had genetic counseling, unpredictable cases of genetic disease would still exist.

Many genetic diseases can be detected at birth by blood tests or other examinations of newborn babies. When applied routinely, such tests are a form of *genetic screening*. Infants who are found to have a genetic disease can sometimes be given treatment. But getting this treatment to those who need it is a problem that raises many hard questions.

For example, there is a genetic disease called phenylketonuria, or PKU for short, that occurs in about one out of 15,000 babies. PKU causes severe mental retardation because a child with the disease cannot digest food properly at the time its brain is developing. But if the disease is detected soon enough, the infant can be placed on a special six-year diet that will prevent most of the damage. Testing for PKU is now done routinely in most American hospitals, and is an example of genetic screening usually thought to benefit everyone. The child with PKU is saved from a life of mental incapacity; society also benefits because it is less expensive to do the screening and to treat PKU victims than to care for mentally retarded persons. Treated PKU patients are not completely free from the disease, however, and if they become mothers, their babies are usually harmed during pregnancy. So screening does not fully solve the PKU problem. In fact, from the standpoint of eugenics, it increases the number of future PKU cases because treatment enables victims of the disease to live normal lives and therefore to marry. A more serious problem is that the tests given at birth tell only which babies show signs of possible PKU, and those in whom the signs turn out to mean something else can suffer harm if they are restricted to the special PKU diet.

There are other diseases, many of which are much

rarer than PKU, that can be detected by genetic screening at birth. Establishment of screening programs to find and treat cases of all these diseases would help their victims. However, screening alone is not sufficient for accurate diagnosis; it merely identifies babies that should be examined by doctors with special knowledge of genetic diseases. There are relatively few such babies, and even fewer prove to have conditions that need treatment. Some people feel the effort and money spent on screening programs would help more people if it were used for other medical projects. The question of whether or not to screen for rare genetic diseases is becoming more controversial as scientists learn to detect signs of more and more of them.

The problem of cost in relation to benefit also arises with diseases that occur mainly in certain ethnic groups. Blacks almost never have PKU, but they do sometimes have sickle-cell anemia, which is very rare in whites. Tay-Sachs disease is hardly ever found outside Jewish families. Little benefit can be obtained by screening members of an ethnic group for diseases that scarcely ever appear in that group, yet screening different people for different things would strike many as discriminatory. Some say the resources spent in screening for diseases that affect only minority groups could be better used in finding cures for more common diseases that occur in all ethnic groups. Such a policy would be even more discriminatory, however.

Economic considerations are not the only arguments against screening programs. Many people believe screening for diseases that cannot yet be treated does more harm than good, especially if they are diseases that don't affect health in childhood. Those discovered to be afflicted can only worry. Then too, when a child is found to have a genetic disease, the parents are identified as carriers, and some people feel this is the same as being labeled "inferior." Although there is no scientific basis for this feeling, it exists, and many people are concerned about the emotional and social problems it has begun to cause. Others fear that records of genetic screening tests will be misused by poorly informed authorities to deny people rights having nothing to do with producing children. There have been cases where this may have happened: some insurance companies, for example, have been accused of discriminating against carriers of sickle-cell anemia although carriers themselves are not diseased.

The pros and cons of genetic screening are often considered part of the controversy about genetic engineering. However, it is important to realize that most of the issues connected with genetic screening of babies have no special relationship to genetics. Such screening is simply a form of medical diagnosis. There are screening programs for many illnesses that are not thought to be genetic: certain forms of cancer, for instance. In America, few of these programs are compulsory; people are not forced to have medical tests except in special instances where contagious diseases are involved. Many people feel that the right to choose whether or not to seek such tests is a basic individual right. According to this view, laws requiring genetic screening are no more justifiable than laws requiring people to have regular medical checkups would be. For this and other reasons, more and more people are coming to feel that parents should be the ones to decide about having their babies tested for genetic diseases. And almost everyone agrees that the results of such tests should be kept as confidential as other medical records.

If society were to adopt a policy of eugenics, genetic screening of prospective parents, rather than of babies, might be made compulsory. All the controversial issues of eugenics would be involved in genetic screening of that type. But at present, medical practices based on knowledge of genetics are not eugenic practices. They are simply ways of helping people who want their children to be as healthy as possible.

4

Tomorrow's Genetics: Better Use of World Resources

So far, less is known about human genes than about the genes of plants, lower animals and primitive organisms like bacteria. It is easier for scientists to learn about the genes of lower organisms because there are fewer per cell than in humans, and because experiments can be done that couldn't be done with people. Lower organisms grow to maturity much faster than human beings, so that many generations can be observed during experiments. Mating can be carefully controlled. Also, various ways of altering genes can be tried.

One purpose of such work is to gain knowledge that may someday be used to make human beings healthier. But the work has more immediate practical applications, too. Genetic improvement of plants, animals and bacteria can, in itself, help solve world problems. Crossbreeding has been increasing the food supply since human history began, but crossbreeding is no longer the only method of improving genotypes; other techniques are being developed.

Genotype is what makes plants and animals of particular kinds, or *species,* more like each other than like representatives of other species. The effects of genes that produce differences between individuals are easiest to study, but genes also control similarities. For instance, individual cats have different fur colors, and genes determine these differences. But genes also determine the things that make all cats look like cats instead of like dogs.

Plants and animals that can be crossbred are usually of the same species. Sometimes members of two similar species can be crossbred, however, and the resulting offspring are called *hybrids.* Hybrids are often superior to either of the parent species, but they are usually unable to reproduce. For example, a mare and a male donkey can be crossbred to get a mule. A mule cannot be bred to produce more mules, however; the mare and donkey must be crossed again for another mule to be born. Despite this disadvantage, mules have long been important farm animals. And hybridization is an even more important technique in plant breeding, where reproduction can be quite easily controlled.

In nature, species that are not similar cannot be crossbred. If their genotypes are too different, male and female reproductive cells will not unite. Because lions and tigers are closely related species, it is possible to crossbreed them and get hybrids; but there is no such thing as a cross between a cat and a dog. Even more obviously, there is no such thing as a natural cross between an animal and a plant. Sexual reproduction requires similarity of genotype.

However, scientists are learning to combine cells from different species in laboratories. This is a useful experimental technique for studying the chemical structure of genes, even though the cells do not grow into creatures.

In the case of plants, it may be possible to make practical use of this cell fusion technique. Hybrid plants that grow and reproduce have been created this way experimentally. Cells taken from two species of tobacco plant have been caused, by special chemicals, to join and become a single cell containing all the genetic information that was in both original cells. The combined cell multiplied and grew into a tiny shoot with leaves, which was attached to one of the parent plants. It produced seeds that sprouted into hybrids able to produce seeds in their turn. Perhaps in the future scientists will use this method to develop hybrids of plant species not at all like each other: tomato plants with carrot roots, for instance, or watermelon vines with

These scientists—Dr. Peter S. Carlson, Rosemarie D. Dearing and Dr. Harold H. Smith—are examining the first mature hybrid plant ever grown by combining cells from two different species into a single cell. COURTESY BROOKHAVEN NATIONAL LABORATORY lettuce leaves. Such plants might make more efficient use of farmland than today's crops.

Hybrids developed from combining cells could have other advantages. Much research is being done to develop crop species that need less fertilizer, or less water, than existing ones. Achieving these goals would not only help solve the problem of world hunger; it would help to conserve Earth's resources.

It may be that plants can be made to produce food without using farmland at all. In other experiments, unfertilized tomato flowers have developed into ripe tomatoes after being cut from their plants and placed in chemicals. Cells from soybeans, which are an important source of protein food, have been made to grow protein without ever becoming plants. Possibly, soybean meal can someday be produced in factories from soybean cells without any plants being needed. If this proves practical, there may be no use for exotic new plant species.

Though cells from different animal species have been combined, the combined cells do not grow into hybrid animals and do not often live long; they are used only for special kinds of research. Combination of animal cells with plant cells has also been successfully tried. The resulting cells, which contain genes from both plants and animals, are sometimes called "plantimals." Plantimals do not develop into organisms, but

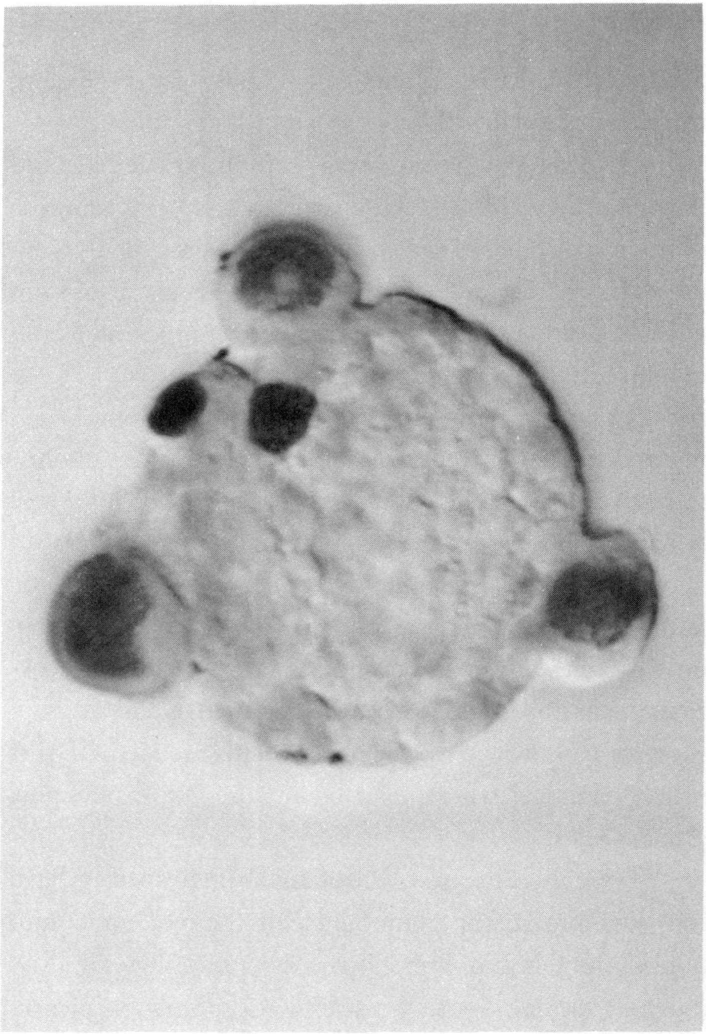

This picture shows a cell made by combining plant cells from tobacco plants and animal cells from humans. Such cells do not grow into creatures. They are used by scientists for learning about plant and animal genes.
COURTESY BROOKHAVEN NATIONAL LABORATORY

they may someday have uses. One idea is that large numbers of them might be grown in factories, just as soybean cells might be grown. These cells would use photosynthesis—the process that plants use to form food—to change light and nutrient chemicals into meatlike substances for people to eat. Such a process could be much more efficient than slaughtering animals for meat. Growing meat in large vats may seem strange today, but population growth could force humanity to take such steps in the future. A century from now, plantimal cells may be quite useful things to have around.

One-celled organisms have been useful throughout history. Bacteria, for instance, are important in making such foods as cheese, yogurt, vinegar and wine. Recently bacteria have been modified by genetic engineering techniques to do entirely new jobs. One way this has been done is through the transfer of *plasmids:* tiny rings inside bacteria that carry genes on them. The genes on plasmids are not needed for the bacteria that have them to live, as other genes are. However, plasmid genes often give a bacterium advantages others of its kind do not have. Some plasmids can cause tiny tubes to form through which they, and other plasmids, can pass from one bacterium to another. By controlling this passage of plasmids between bacterial strains, scientists can place specific genes in specific bacteria, giving these bacteria new abilities.

For example, there are plasmids that enable bacteria to eat oil in addition to their regular food. One scientist took four different plasmids, each of which allows a bacterium to eat a different component of oil, and used the natural ability of plasmids to pass between bacteria to get all four of these plasmids into one bacterium. Because they were all duplicated when the bacterium reproduced, this resulted in a new strain of bacteria that can digest many kinds of oil fast and efficiently. Oil is sometimes spilled into the ocean by accidents on offshore drilling platforms or supertanker ships; the new bacteria may prove very helpful in cleaning up future oil spills. However, this cannot be tried until scientists are sure that it is safe to release the new bacterial strains into the environment.

In the future, bacteria may be used to extract metals from low-grade ores, perhaps even from seawater. The high-grade ores from which metals are extracted by traditional mining methods are becoming scarce. However, small quantities of metals exist in many plentiful substances. Species of bacteria that can extract metals from mining wastes were discovered some years ago; they are already being used in many countries, especially for producing copper and gold. By controlling transfer of plasmids between these species, it may become possible to develop bacteria that can do this job much faster than existing ones can. It may also be pos-

Dr. Ananda M. Chakrabarty has developed a strain of bacteria that can eat oil faster than any bacteria found in nature can. He combined genes from four different strains to do this. The spots on the disk in this picture are colonies of the new bacteria. The two flasks contain oil and water; bacteria have begun to digest the oil in the one at the right.

COURTESY GENERAL ELECTRIC RESEARCH AND DEVELOPMENT CENTER

sible to get metals from sources such as seawater that have not previously been used as "mines."

There are many other potential applications for bacterial processing and similar operations. It has been suggested, for example, that a certain type of fungus could convert waste material into a form of fuel. Almost everyone agrees that it would be desirable to recycle all kinds of wastes instead of throwing them away. Using genetic engineering to make organisms do such recycling could prove to be a major step forward.

All these ideas for modifying organisms may bring great benefits to civilization. But many scientists warn that not enough is yet known to use altered organisms; there is no way to be sure that organisms modified to do one job won't also do things that cannot be foreseen. A new strain of bacteria could develop dangerous characteristics in addition to the characteristics it was designed for. There is still a great deal to be learned about genetics before anyone can evaluate the risks.

Meanwhile, scientists are continuing to develop ways of producing more food that don't depend on designing new organisms. Besides crossbreeding to get cattle that produce more milk and meat, for instance, they are working on ways to produce more such cattle. Calves fathered by prize bulls can already be obtained in greater quantity through artificial insemination than through trying to keep a prize bull available to mate with large numbers of cows. But just as some bulls are of more desirable genotypes than others, so are some cows, and cows rarely have more than one calf at a time. Also, because of laws to prevent the spread of cattle diseases, cows can't be taken from country to country in all cases, when farmers want to import superior breeds. So veterinarians have developed a new method called *ovum transplant* that allows genetically desirable cows to produce more calves. A cow can be given drugs that cause her ovaries to release many eggs at a time. These eggs are fertilized by artificial insemination. A few days later, the embryos are removed from her womb and implanted in other cows, where they grow into normal calves just as if they had resulted from ordinary mating. This way, many cows can give birth to offspring whose "real" or genetic parents have especially fine characteristics.

Obtaining superior calves by ovum transplants is an expensive process, however; and neither that technique nor artificial insemination can be used as successfully on other livestock as on cattle. Many scientists believe that it will eventually be possible to reproduce the best animals—and also the best plants—by *cloning*. This technique would bypass sexual reproduction entirely and result in "offspring" genetically identical to a single "parent" organism.

In cloning, scientists use microsurgery to remove

the genes from almost any cell, except a reproductive cell, of the animal or plant to be duplicated. An egg cell from the same species of organism is exposed to radiation, which destroys the egg cell's genes without seriously damaging the cell itself. The genes from the first cell are then placed in the egg cell, and the egg is made to grow. It has all the genes needed to grow into a complete new organism no matter what type of cell these genes originally came from. Since they are the ones possessed by the chosen "parent" organism, an identical twin of the "parent" is formed. There is no known biological limit to the number of such twins that can be cloned from a single organism.

Experimental cloned copies have been made of relatively simple organisms such as frogs and carrots, but cloning may well be possible for more complex ones as well. This would have important practical uses. Superior livestock could be cloned by farmers. Prize-winning flowers or race horses could be cloned to provide exact replacements when the originals died. Identical laboratory rats and guinea pigs produced by cloning would be of great use in medical research. Cells from the bodies of endangered species of plants and animals could be frozen or placed in tissue cultures; these species could be preserved by cloning if they became extinct in their natural environments.

However, cloning could have disadvantages if

used too widely. Some scientists fear that it might be used so much that valuable genes would be forever lost. No single organism has all the desirable genes of its species. In nature, sexual reproduction is a means of preserving variety in what is called a species' *gene pool.* If environmental conditions change, or if a new disease strikes, variety is important. One genotype may prove less able to survive than some slightly different genotype. "Superior" plants and animals can be chosen for specific uses at specific times, but no one can know what characteristics will prove "superior" in the future. Scientists are concerned about this problem even now when crops and livestock are being made more and more uniform simply by controlled crossbreeding. Cloning could create a more serious problem.

Selection of genotype, however it is accomplished, is an essential part of agriculture. New, efficient techniques are badly needed to cope with hunger and depletion of Earth's resources. But these techniques must be wisely used if life on Earth is to remain adaptable to whatever the future may bring.

5

Tomorrow's Genetics: Better Medical Care

Genetic diseases are caused by defective genes in the cells of a human being. Many of these illnesses can be diagnosed today, but only a few can be treated. Even those that can be treated cannot be cured. Treatment eliminates some, or occasionally all, of a disease's symptoms; it makes it possible for a patient to live more nearly as people without illness do. But in most cases treatment must be continued for the rest of the patient's life. If it is stopped, he or she will be sick again, or perhaps die. A cure, on the other hand, would get rid of the disease entirely, so that no more treatment of its victim would be needed. Someday genetic engineering may be used to cure genetic diseases. Scientists are already working toward this goal.

One way of curing a genetic disease would be to

replace the defective genes with working man-made ones. After all, genes are chemicals, and chemicals have been manufactured for hundreds of years. Genes are much more complicated than other chemicals, however; synthesizing them is not at all easy. So far the specific chemical structure of most human genes is not even known. It will be a long time before doctors can replace genes in human bodies, but many scientists believe such cures will eventually become possible. Many people, in fact, believe this is the most important potential benefit of research in genetics.

But there are conflicting views. For instance, some feel it will never be practical to cure disease by modifying genes. Others feel that not enough people could be helped to make it worth the cost and dangers of developing a capability to alter human genetic material. If scientists had such an ability, they say, it could be misused. People could be changed in ways that had nothing to do with curing disease. However, not everyone believes the danger of society's misusing this capability would be any greater than the danger of its misusing of already existing medical capabilities.

Modification of genes for the purpose of curing disease is called *genetic therapy*. The difficulty of synthesizing genes is by no means the only problem to be solved before genetic therapy can be used by doctors. There is also the question of how to get them into the cells of the patient's body and how to make them function there. Several ways this might be done have been proposed.

Genes can be inserted into cells by *viruses*. A virus is not much more than a few genes surrounded by a chemical coating. Scientists are not sure whether viruses are living or nonliving things. They cannot reproduce on their own, which is one of the most important characteristics of life. Instead, they attach themselves to a cell of another organism and inject their genes into it. The parts of the cell responsible for reproduction are then forced to make more viruses using the invading virus's genes.

Cells invaded by viruses often die; indeed, viruses are the cause of many sicknesses, including the common cold. Scientists think some viruses may cause cancer. But other viruses might be useful. It is possible to use them to insert genes into bacteria, and so perhaps they could be used to insert genes into human cells to replace defective ones. This kind of virus invasion would not kill the cells; instead, it would add needed genes to the cells' genetic makeup. However, not enough is yet known to be sure that harmful genes wouldn't be inserted along with the needed ones.

Scientists have already tried to cure one genetic disease with virus injections. A virus was discovered that contains the genetic information for producing a substance called arginase. Human beings need arginase, and those who cannot produce it suffer from a rare disease that grows progressively worse, causing brain damage and many other serious symptoms. A doctor injected the arginase-producing virus into three children who have this disease. Unfortunately it did not help them; scientists are not sure why, but it may have been because the virus had been stored too long, or because the children were already too old to be cured. Some people believe this experiment should not have been done because there was no way to know that the virus couldn't cause harm. Other people feel that since the children were already sick and were becoming mentally retarded, they couldn't have been made any worse off, and it was worth taking a risk.

Another way of inserting functioning genes into a diseased body would be through transplants of organs or tissues. Such transplants can cure some diseases, and they might cure many others, both inherited genetic ones and those caused by damage to the body—or possibly to genes—during life. But most types of transplants are difficult if not impossible now because the human body rejects cells not its own. This is called the *immune response:* a person's body is ordinarily immune to material not originally part of it.

The immune response is a protection against harmful invasion, and like all other physical processes of the

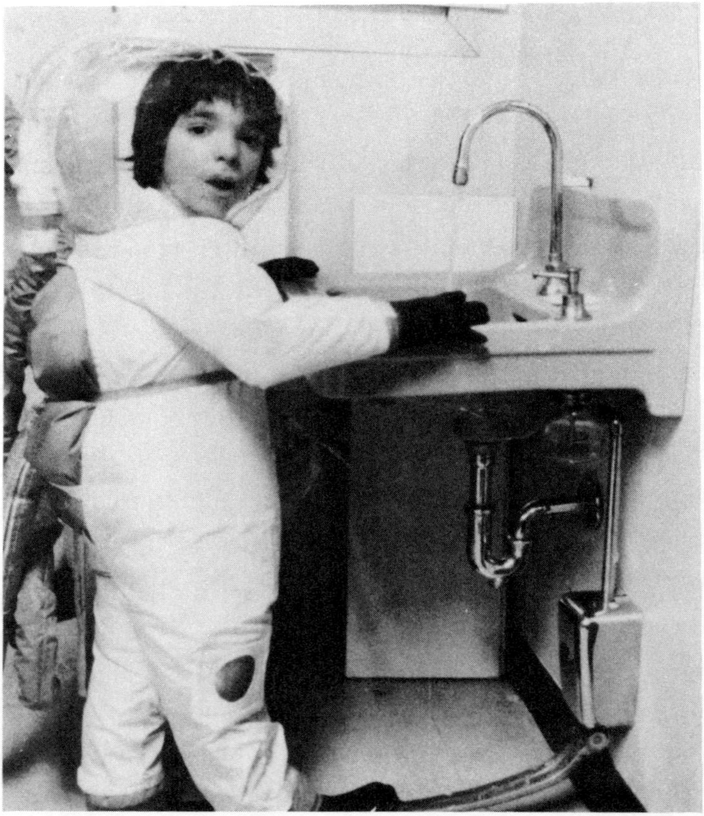

David, who is 6 years old, was born with a genetic defect that causes him to lack immune protection against disease. Doctors can treat this defect by confining him to a special germ-free plastic bubble and sterilizing his food, clothes, toys and books. They hope that someday a way to cure him will be found. Meanwhile he can leave his "bubble" only by wearing a space suit made for him by NASA. In this picture, David—safely enclosed in his space suit—is trying out a pedal-operated water faucet; but he cannot touch the water. COURTESY TEXAS CHILDREN'S HOSPITAL body, it is controlled by genes. Children born with a genetic defect that causes them to lack an immune system usually die, although ways of treating them are being developed. Doctors are beginning to believe that some diseases that strike later in life, such as cancer, may also be the result of deficiencies in the immune system. Other diseases are caused by the body becoming immune to some of its own tissues by mistake. If enough can be learned about genetic control of immunity, perhaps all these types of diseases can be cured; and perhaps transplants can be accomplished more easily.

At the present time, transplants are possible only between close relatives who have very similar genotypes, or through the use of drugs that temporarily suppress the immune response. These methods do not always work. A much safer and more successful means of transplant therapy would be to grow the organs or tissues in laboratories from the patient's own cells. Since these cells would have a genotype identical to the person's other cells, no immune response would occur. Theoretically, it should be possible to grow any type of tissue from any cell because every cell contains all the genes for the whole body; but no one yet knows how to do this. Before it can be done scientists must learn more about what turns specific genes on and off.

Therefore, although many scientists and physicians consider genetic therapy an extremely promising field for medical research, the goal of using it to cure disease cannot be attained quickly. Most scientists don't talk about potential cures because they do not want to give the impression that they're making promises. For this reason, people who have heard of genetic engineering are not always aware that medical benefits are among its long-term goals. They are more apt to hear about the sorts of aims mentioned in science fiction, such as giving people gills so they can breathe under water. Actually, geneticists are far more interested in curing illness than in redesigning human bodies to include gills.

Learning how to get functioning genes into patients' bodies is not all that is needed to cure genetic diseases, however. Each human cell contains hundreds of thousands of genes, relatively few of which function in that particular cell; and all of these genes must work together in exactly the right way within the body. Scientists do not understand nearly enough about how they work to start modifying them. They cannot be sure that fixing one defective gene wouldn't cause others to malfunction. They do not yet know the specific function of most human genes. Some scientists feel it would not be safe to try repairing genetic defects without first learning all about the interactions between all the genes humans possess. Others feel the risks of doing so would be no greater than they are in other types of medical experiments, some of which may have been affecting genes all along without anyone's realizing it. Almost everybody agrees, however, that genetic therapy must wait until more is known about the control of genes than present experiments will show.

Long before genetic therapy can be developed, other forms of genetic engineering may be of great practical benefit to people with genetic diseases. For instance, diabetes is a relatively common disease thought to be at least partly due to defective genes. Many people who have diabetes need treatment with the drug insulin, either because their bodies don't produce enough of it or because their cells don't respond well to it. Before the discovery of insulin, patients with severe types of diabetes usually died; now they are able to live normally with daily injections. But insulin is obtained from cattle and pigs, and the demand for it is becoming greater than the supply. A more economical way of obtaining insulin would be to make it in laboratories through the use of bacteria that have been given genes for producing it. Scientists are actively working toward this. In 1977, they succeeded in inserting the gene that produces insulin in rats into bacteria, and they may soon be able to insert human insulin genes into bacteria. Someday, bacteria may be used to produce many kinds of drugs.

There are people who believe that if it should become possible to cure genetic diseases rather than merely to treat them, this would not be good for the human race. They feel curing such diseases would simply add to a problem that has already arisen with treatment: the number of defective genes in the human race increases when people with genetic defects become able to marry and have children. Most forms of genetic therapy scientists have proposed would cure only the individual patient; the therapy would not repair the genes an afflicted person might pass on to offspring. Yet patients wouldn't be apt to worry about having children if they knew these children, too, could be cured. The people who do worry believe that this would lead to dependence of the human race on medical technology, which they feel would be bad.

For this reason, these people think that instead of trying to cure genetic diseases, society should try to prevent babies with such diseases from being born. This could be done through eugenics programs. It could also be done by simply not offering cures, though few would approve of not using medical therapy that is available. The question is whether the development of new therapies is a desirable goal. Some argue that the goal in genetics should be the modification of genotypes of unborn children rather than the repair of defects after birth.

Whether or not genetic therapy is developed to repair defects, it may someday be possible to control babies' initial genotypes. Some scientists, for instance, think that it may become feasible to modify genes in human sperm sooner than genes within the human body. This alone would not solve all the problems of genetic disease, but it might help. And perhaps, in time, genetic therapy can repair the patient's reproductive cells as well as whatever cells are malfunctioning, so that any future children of that patient would be born with undefective genes.

It is important to realize that genetic therapy could also be applied to diseases other than inherited ones. For example, although cancer and heart disease are not in themselves inherited, many scientists believe that the genotypes of some people may make them more likely than other people to get these diseases under certain circumstances. And even cases caused by environmental factors alone may result from damage to the genes of certain cells. No one yet knows the real reason some people get illnesses that others don't. But people who favor genetic therapy believe that it will help victims of common diseases as well as rare ones.

6

A New Era in Genetic Research

Scientists have studied genetics for many years, but until the middle of this century, they were able to study no more than the effects of genes on heredity. They could observe only the traits that appeared in plants and animals after controlled crossbreeding. No one knew the chemical structure of genes. In the 1950s, however, scientists found that genes are made of a chemical called DNA. Since then, a great deal has been learned about DNA. A whole new field of study has developed called *molecular genetics* because it deals with DNA molecules consisting of genes, rather than with traits genes produce in organisms.

Many people believe that what molecular geneticists are doing in the 1970s is the most significant scientific development since the release of atomic energy—

and in the eyes of some, it is more dangerous. Others feel its potential for good is far greater than its dangers. The reason for all the excitement is not merely the new knowledge of genes. It is the development of a new type of experiment called a *recombinant DNA experiment.*

Scientists once thought that when they understood how genes in bacteria work, they could figure out how genes work in higher organisms. But they found that wasn't true. Higher organisms have many, many more genes than bacteria, and these genes aren't controlled in exactly the same way as bacterial genes. And it seemed impossible to isolate such genes to study them. Many scientists were discouraged; they were afraid they could never learn how the genes of higher organisms work.

Then a technique for isolating genes was developed: the recombinant DNA experiment, which is also known as gene splicing. DNA molecules are double strands or chains of special chemical "building blocks" that "spell out" genes just as the letters of the alphabet are building blocks that spell out words. Genes are strung together one after another on the DNA chains, and most genes in higher organisms consist of more than a thousand building blocks each. In recombinant DNA or gene splicing experiments, DNA chains can be cut into individual genes or small groups of genes. This makes them easier to study. Furthermore, the gene

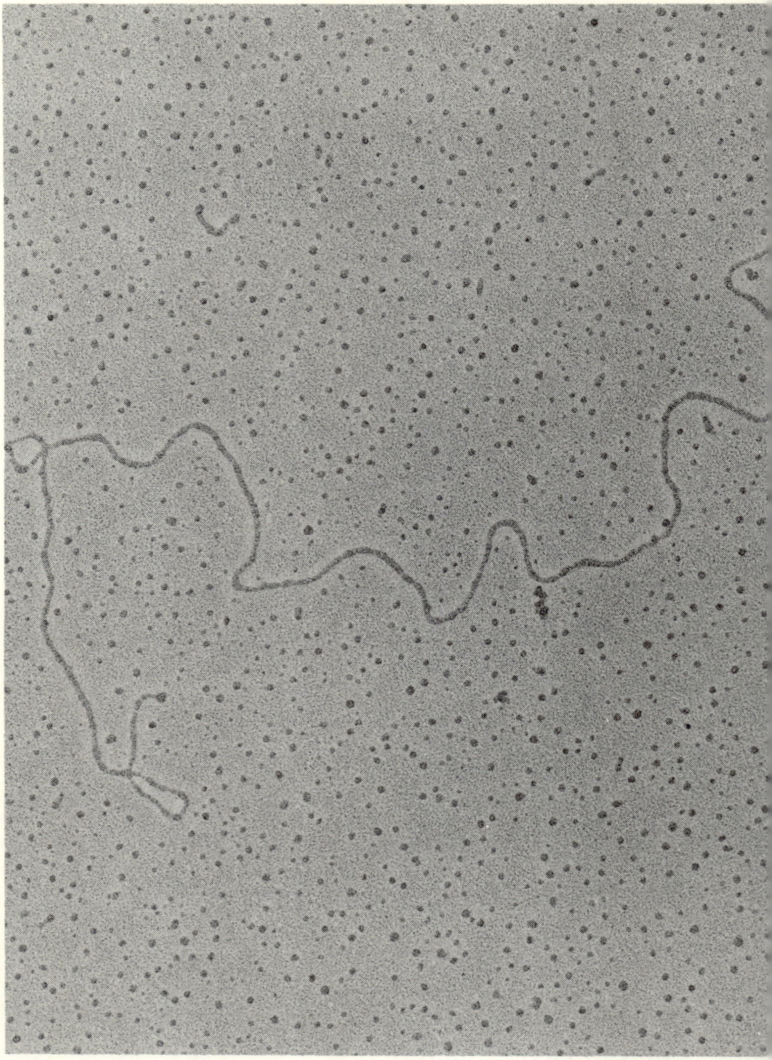

This picture was taken by a very powerful electron microscope. It shows a DNA molecule composed of about 25 genes. Even with a microscope, it is not possible to see

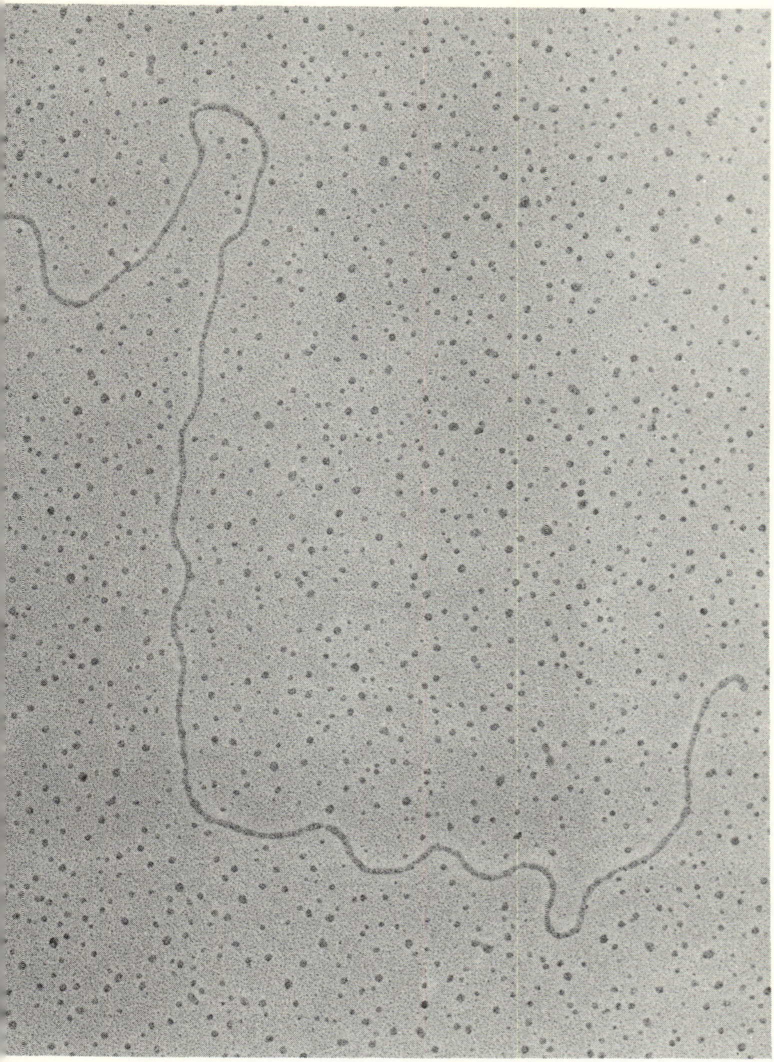

where one gene ends and another begins.

COURTESY BROOKHAVEN NATIONAL LABORATORY

splicing technique offers a way to recombine natural genes with synthesized ones for use in genetic engineering.

The DNA strands are cut into pieces during experiments by a chemical called a *restriction enzyme.* If the cut-up DNA is then mixed with DNA from a different organism that has been cut up by the same restriction enzyme, the two kinds of DNA recombine into one double chain. Genes of the first organism are intermingled, on this new chain, with genes of the second. A new combination of genes not found in nature is produced.

Of course, this in itself does not make a new kind of creature; it does not make any creature at all. DNA is not in itself alive. But recombinant DNA made by combining plasmids or viruses with genes from other organisms can be inserted into bacteria or yeasts. These one-celled organisms are alive, and they multiply. The new genes multiply with the original genes of the organism. Some people consider such modified organisms "new creatures" and other don't, but whatever words are used, the organisms have new abilities if the spliced-in genes function.

Experiments of this kind are an extremely valuable research tool. One gene is much too small for scientists to analyze; they need many copies of the same gene. Before gene splicing methods became available it was not

Microorganisms must be grown under carefully controlled conditions in biological experiments, and Petri dishes are often used for this purpose. Bacteria placed on the end of the metal probe are transferred to the nutrient gel in the Petri dish. A clear lid is then placed over the dish to prevent other bacteria in the air from growing there as well.
COURTESY NATIONAL INSTITUTES OF HEALTH

possible to get enough of any particular kind of plant or animal gene to study, even if that gene could have been separated from all the hundreds of thousands of others in a cell. Now scientists can do both: they can cut up genetic material, and they can duplicate the genes they want to study by putting them into one-celled organisms. Most molecular geneticists feel recombinant DNA experiments will prove to be the key to learning how genes work and to decoding the genetic structure of plants and animals. Already these experiments have shown that the structure of animal genes is not as much like that of bacterial genes as had been thought.

In most cases, scientists cannot yet plan what specific genes they will attach to plasmids; they must do what are called "shotgun" experiments, cutting up DNA and hoping to get a gene that will show something of importance. But in other instances they are able to splice in man-made copies of genes chosen in advance. Eventually, they may be able to splice in specifically chosen genes from cut-up natural DNA, also.

It was through recombinant DNA experiments that the gene for producing rat insulin was put into bacteria. At first the bacteria did not actually produce any insulin; it was not "turned on." However, in more recent experiments, spliced-in genes have caused bacteria to produce insulin and other chemicals normally produced only by mammals. Soon it may be possible to make bacteria manufacture many substances needed for treatment of people whose own genes are defective. Gene splicing is also a potentially more efficient technique than cell fusion or natural plasmid transfer for designing bacteria to do other kinds of special jobs.

This does not sound dangerous, or even very controversial, to most people. Why should anyone oppose recombinant DNA experiments if they are so vital for learning about genes, and so useful for putting lower organisms to work in fields like medicine and agriculture?

One reason is that recombinant DNA technology is a key not only to understanding genetic makeup, but to altering it—in higher organisms as well as lower ones. Some people feel that human beings should not change species from the ways they have evolved in nature. To be sure, species have been altered throughout human history by means of controlled crossbreeding; but sexual reproduction has built-in safeguards that prevent alterations from being too drastic. Recombinant DNA technology can bypass these natural safeguards and permit scientists to develop entirely new kinds of creatures. Some say this would be "creating new forms of life," while others point out that it is not "creation" to change living things. Even recombinant DNA methods cannot make non-living matter live. Nevertheless, there are people who feel that making major changes in higher organisms, and especially in the human race, would be "playing God." In the eyes of some, this is wrong, no matter how good the results are. Others feel it is not wrong in itself but that human beings do not yet have the wisdom to predict the results and are not likely to gain enough wisdom in time to prevent harm to Earth's ecology.

No one worried much about such harm being done by the study of genetics until recombinant DNA experiments began. Making new genes did not seem practical enough to be considered dangerous. A group of scientists worked for nine years to make the first functioning man-made gene, and it was a fairly simple one, a copy of an existing bacterial gene. It had far fewer building blocks than human genes; synthesizing human genes seemed to lie far in the future. Furthermore, scientists weren't sure that a gene from an unrelated organism could be made to function in a bacterial cell, even if it could be added to bacterial DNA by a plasmid or virus.

But many developments occurred in a short period of time. Even before synthesization of the functioning bacterial gene was complete, gene splicing offered an easy way to put new genes into plasmids and viruses. In addition, scientists found ways to make genes without analyzing and copying the sequence of their chemical building blocks. One such method allows them to copy genes without knowing that sequence.

Genes, which are made of DNA, work by means of another chemical called *messenger RNA*, which is used by cells to carry out the genes' instructions. Copies of the functioning genes in each cell are made of messenger RNA. Scientists can isolate the messenger RNA molecules more easily than they can isolate genes, and it is possible to reverse the normal copying process, so that these molecules can be used to produce genes like the original ones. This was the method used for getting the rat insulin genes used in experiments.

It is also possible to synthesize genes by figuring out the sequence of their building blocks from the chemical structure of the substance they produce. This is a shortcut because substances produced by mammals can, in many cases, be isolated more easily than either genes or messenger RNA molecules. The first non-bacterial gene scientists got to function in bacteria was obtained by this method. It enabled the bacteria to make a chemical normally made in the brains of mammals, and the experiment proved that spliced-in genes can be turned on whether or not they are naturally related to the organism in which they are placed.

Copying a gene that already belongs to bacteria is one thing; putting a gene taken or copied from another creature into bacteria is something else. To some people it is exciting; to others it is frightening. The fact that modification of bacteria may be a step toward modification of higher organisms is only one of the disturbing possibilities. Bacteria themselves affect higher organisms; they are a major cause of disease. Dangerous genes could be transplanted to bacteria just as easily as helpful ones. In fact, dangerous characteristics now possessed only by rare bacteria could be given to common strains.

Thus, a second major reason many people have

worries about recombinant DNA experiments is that these experiments are in themselves dangerous if they are not carefully controlled. For example, there are viruses known to cause tumors in animals. Putting genes from these viruses into bacteria would enable scientists to learn more about what causes human tumors and might lead to knowledge about how to prevent cancer or even cure it. But if the bacteria given such genes were to accidentally escape from a laboratory, they might infect human beings. Of course, medical researchers have always had to do experiments with disease-causing organisms, but most such organisms cannot survive in ordinary environments long enough to cause widespread disasters. Recombinant DNA experiments might produce some that could.

Another hazard of gene splicing is the possibility that genes that make bacteria resistant to antibiotic drugs could be transferred to unresistant strains, so that drugs would no longer work against the diseases they now cure. Already many bacteria are resistant to such drugs, and more of them become resistant by natural means as time passes. This is a serious medical problem. In recombinant DNA experiments, resistance of bacteria to specific drugs is often used as a way of telling which of the bacteria have new genes and which don't. If strains not already resistant to an antibiotic drug were given resistance, and they escaped, the problem of drugs becoming ineffective could grow worse.

Even gene-splicing experiments that seem harmless might have hidden threats. Some scientists have suggested that two harmless genes from different organisms might form a dangerous combination when placed together in a single organism. With today's limited knowledge, no one can tell whether or not this could happen. All splicing of genes that aren't thoroughly understood carries the risk, however slight, of producing potentially harmful microorganisms.

Soon after the invention of the recombinant DNA technique, many scientists realized that it could be dangerous. They agreed to stop the most hazardous kinds of experiments completely for six months until safety guidelines could be established. This was a very unusual thing to do. Ordinarily, scientists want their work to continue uninterrupted, and they rarely feel there is good reason for interruption. Though some scientists in the past had felt the results of their work could be dangerous if misused, there had not been a previous case in which the research itself was risky to anyone except the people doing it. But gene splicing is different. Scientists saw that misuse of new knowledge was not the only thing to worry about. Unless precautions were taken, accidents might happen during the actual experiments that could affect people outside laboratories as well as the experimenters.

Early in 1975, safety rules were developed for recombinant DNA work that most scientists around the world agreed to follow. This meant banning the riskiest types of experiments and doing certain other types in specially equipped laboratories from which it is practically impossible for bacteria to escape. It also meant using bacteria that can't survive outside laboratories for some kinds of experiments. Since such organisms obviously do not exist in nature, scientists produced them using genetic engineering techniques. These special bacteria cannot live without chemicals that are supplied to them in labs; furthermore, they die at the body temperatures of humans and other mammals, so they cannot infect people. If one did manage to escape from a lab, it would have less than one chance in 100 million of surviving.

Not all scientists approve of the details of these rules. Some think they are too strict, although most agree it is better to be safe than sorry. Others believe they are not strict enough. People who are not scientists have also become involved in the debate about the rules, and in some places voluntary rules have been replaced with laws. It is likely that particular safety standards will change as more and more is learned about gene splicing.

There are some people, including some scientists, who believe recombinant DNA research is just too dangerous to be done at all. Others believe all hazardous gene splicing experiments should be done in space, in-

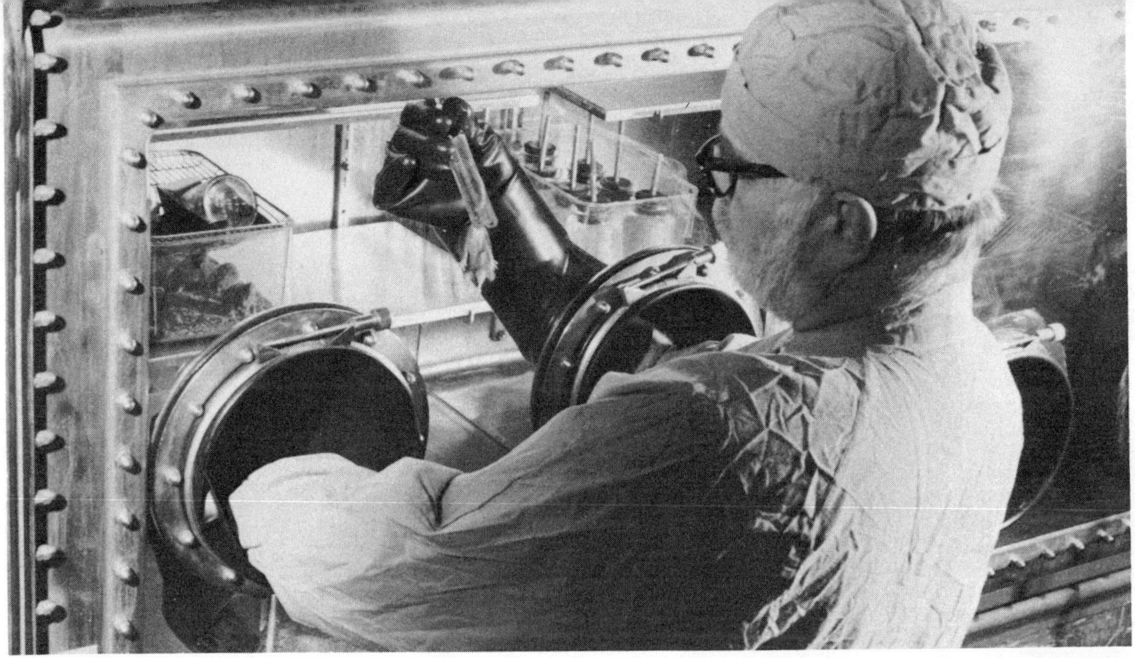

In specially-equipped laboratories, scientists doing types of recombinant DNA *experiments that might be dangerous do not let their hands touch tools or materials. Everything is kept inside sealed cabinets called glove boxes, which have shoulder-length heavy gloves tightly attached to their openings. This laboratory worker is reaching into such a cabinet by putting his arms into the gloves. The front of the cabinet is made of glass so that he can see the things he is handling.*
COURTESY NATIONAL INSTITUTES OF HEALTH

side orbiting laboratories, where there would be no chance of any altered organisms getting loose on Earth. The majority of people, however, do not consider it necessary either to completely stop recombinant DNA research or to wait until there are genetics labs in orbit. They believe the potential benefits of such research make it worth pursuing, as long as safety precautions are carefully observed.

7

Some Ideas for the Distant Future

A great deal of what is being said about genetic engineering today concerns not just new methods in agriculture and new therapy for disease, but new ways of making babies. There is also quite a bit of talk about modifying genes to give people new abilities and perhaps far longer lifespans. Therefore, to many people, the term "genetic engineering" means altering some of the most fundamental facts of human biology. On the whole, opponents of experimentation in genetics are more apt to use this narrow definition than supporters. But whether or not they favor the experimentation, most people feel it may eventually lead to physical changes in human beings beyond the cure of illness.

This is not something likely to happen soon. It is not the immediate goal of scientists now working in genetics; and even if it were, it would be a goal that could not be attained until much more has been learned about genes. However, the work being done at present may be the first step toward far-reaching changes in human reproduction and in the human body's capabilities.

Even now, new ways of making babies are being developed. This is rarely a genetic engineering technique at present. It usually has nothing to do with genes; it is simply a medical procedure for helping childless couples who want babies of their own. Artificial insemination is already chosen by some such couples to enable the wife to become pregnant. A very few experiments have also been done with fertilizing the wife's ovum outside her body and then implanting it, as a method of overcoming certain types of infertility.

However, not all couples with an infertility problem can be helped by these methods. Although many are happy to adopt babies, others have a strong desire to be the genetic parents of their children. Some doctors think a means of transplanting embryos from one woman to another, or even of growing babies in artificial wombs, would be beneficial because it would bring happiness to these couples. The majority of people feel differently; to them, it seems too far from what is natural. Artificial wombs, in particular, are an idea many people find disturbing. No one knows whether a baby could develop normally away from a mother's body. Probably the physical problems could be solved in time through experiments on animals. But psychological problems might occur in human babies that would not arise in animal experiments, and almost everyone agrees that ethical problems would result from experimentation on babies.

There are people who believe that the feeling against artificial wombs will change. They point out that there would be advantages to starting some babies, at least, outside their mothers' bodies, even if the embryos were implanted in a mother at an early stage. For instance, genetic therapy could cure diseases in embryos more easily than in fully developed babies, and it would be more practical to perform such therapy in laboratories than in women's bodies. On the other hand, some believe that if women could have babies without being pregnant, few would want to bear children; use of artificial wombs or "host mothers" might become common. Some people say that in the distant future, a complete separation of having babies from making love won't seem unnatural. Families, they argue, will not be like families of today. A baby may no longer be viewed as a personification of the love between its father and mother. Perhaps couples will no longer care about being the genetic parents of the children they rear; perhaps they will prefer to have babies started from reproductive cells of especially famous or talented men and women. Ideas like this are frequently discussed. They raise many difficult and complex questions.

Some of the questions have little connection with genetic engineering. Many people, for instance, believe that it would be extremely bad for civilization if the structure of families or the relationship between parents and children were to change. It would be bad, according to this view, whether genetic changes were involved or not.

But other questions do center on genetics. Although artificial insemination is not a genetic engineering technique when it is used because of infertility, it is a form of eugenics when used because the husband has defective genes he does not want to pass on. If it became customary for couples to choose the genetic parents of their children instead of being the genetic parents themselves, that would be widespread eugenics. And if governments were to regulate who was, or was not, permitted to become a genetic parent, that would be a compulsory eugenics program that today would be considered a violation of individual rights. Some people feel that such a eugenics program would be desirable because it would improve the human gene pool. Others feel that individual rights are more important than controlling the gene pool, especially since individuals have different opinions about which genotypes are "best."

Then too, efforts to choose human genotypes—whether individual or governmental—might not really improve the gene pool. In the first place, genotypes might be chosen according to unproven theories, or even current fashions. People might want admired leaders or sports stars or entertainers to be the genetic parents of their children, yet by the time the children were grown, different characteristics might be the ones admired. Furthermore, it is still very uncertain which admired talents are passed on by genes. And it is not sure which "defective" genes are wholly "bad"; even some that now do only harm might turn out to provide enough advantages in certain environments to be worth preserving. Perhaps the greatest danger in controlling the gene pool of the human race is that its variety would be reduced. People would tend to become more and more alike physically—and possibly, in the opinion of some scientists, mentally. It is already known that reducing the variety of plant and animal genotypes may be risky, even when only traditional crossbreeding methods are used. Although human variety could not be reduced as much, any elimination of genes, other than those whose only effect is to produce serious disease in youth, might be harmful in the long run.

Even greater loss of variety might occur if human beings were reproduced by cloning Just as plants and animals can be cloned so that many identical twins of a "parent" organism can be obtained without sexual reproduction, it may become possible to clone humans. This is not just a science fiction idea; some distinguished scientists have discussed it.

Some say it would be beneficial. It's been suggested that "copies" of great men and women could contribute a lot to society, or that single people might wish to have children of their own without a second parent. Others feel it would be a great mistake to produce cloned persons. A cloned baby would take as long to develop and mature as any other; cloning couldn't result in "people factories" that would turn out "carbon copy" adults. But a cloned person would be different from other people because in a biological sense, he or she would be less unique. An ordinary baby has only half its father's genes and half its mother's genes; each person who is not an identical twin is an individual, biologically, because these genes have been combined in a new way. Having one twin brother or sister doesn't make much difference. But having many identical twins—maybe hundreds—could make a person feel less like an individual.

On the other hand, cloned individuals would not really be as identical as is often supposed. They would look alike, but their personalities would be no more "identical" than those of natural identical twins. It cannot be assumed that if they were "copies" of famous people they would achieve similar things; it is not yet known what role genotype plays in determining human abilities and choices.

These are only a few of the questions that are raised in connection with new forms of human reproduction. Many people are troubled by such questions and feel that, because of them, research in genetic engineering may not be a desirable thing. They believe the benefits in agriculture and medicine that may come from advances in genetics are not worth the dangers of opening the door to other developments. This is especially true when genetic therapy is considered. If it becomes possible to correct defective genes in unborn babies, it will also be possible to make genetic changes of other types. In the distant future it might be possible for a dictator to give babies the characteristics he wanted them to have; he might make them into slaves or even turn them into copies of himself. But that sort of misuse is just one of the things people fear. Suppose people could be given superior talents or intelligence by genetic engineering. Who would decide what was "superior"? What would happen if most children were superior to their parents? What about "designing" people for special jobs, like colonizing planets with alien environments?

Another possibility is that genetic engineering might lengthen human lifespan. Some scientists think old age is genetically controlled, and that by means of genetic changes, lives could be extended, perhaps by hundreds of years. Would that be good? Many people say that of course it would, but others are not so sure. Earth is already overpopulated. Besides, people might not really wish to live so long when the time came, even if they were healthy; they might get tired and bored. There are some who believe it would be all right for them to end their own lives when they chose, and others who believe that would be wrong.

None of these things need to be decided soon. It will be many, many years before scientists know enough about human genetics to alter human capabilities. But some people believe that the alteration of bacterial capabilities is a step that will inevitably lead in that direction. They feel it is important to think about future possibilities far ahead of time so that mankind can avoid serious mistakes.

A few people, some of whom are scientists, believe that it is unwise even to start changing natural genotypes by any means other than the traditional one of crossbreeding. They are convinced that mistakes cannot be avoided and that the risks therefore outweigh the benefits. They say that although mistakes occur in every area of human progress, those made in connection with genetic engineering would be worse than all the rest because they couldn't be corrected: genetic changes to any species would be passed on to all future generations.

Other people feel differently. They agree that it is necessary to be cautious about what is done with genes, but they believe much good can be achieved and that if people do think about what might happen in the future, irreversible mistakes can be prevented. In any case, they argue, the search for knowledge cannot be stopped. Most would not wish to stop it. But even some who favor stopping believe that human beings will go ahead and learn all that it's possible to learn, whether society approves or not.

And there are those who believe whatever mankind does to "alter" nature is itself a part of nature. To the argument that man-made genetic changes would interfere with natural evolution on Earth, they reply that scientific progress is natural to the human race. It is probably natural to whatever intelligent beings may inhabit other worlds. In this view, the really unnatural thing would be to call a halt to human evolution by not pursuing new discoveries. Since human nature was what made people develop civilization and technology, saying new uses of scientific knowledge are "unnatural" is the same as saying human nature itself violates natural laws. Knowledge about genes is newer than most kinds of knowledge. But like the others, it can be used wisely or unwisely—the choice belongs to people of tomorrow.

Index